THE GRAND UNIFIED FIELD THEORY II

The Search for the Structure of Space and the Universe

SECOND EDITION

BALUNGI FRANCIS

Website: https://www.bit.ly/quantumgravityresearcher

TABLE OF CONTENTS

DEDICATION

To

Sir Roger Penrose, Erik Verlinde, Sabine Hossenfielder, Carlo Rovelli, Lee Smolin and Neil deGrasse Tyson

AUTHOR'S NOTE

It has been a long time to reconcile quantum physics and general relativity. To date, no globally accepted theory has been proposed to explain all physical observations save for the theory herein presented.

The current understanding of gravity is based on Albert Einstein's general theory of relativity, which incorporates his theory of special relativity and deeply modifies the understanding of concepts like time and space. Within GR, gravity has two possible interpretations: a field one and a geometric one. According to the latter, that has become the prevalent one, gravity is due to the curvature of the space - time "tissue", represented as a "rubber sheet", due to the presence of a mass. Nevertheless, this is a purely mathematical description telling nothing about the physical mechanism starting the motion. In fact, even supposing the existence, in the neighbouring of a source mass, of a curved four - dimensional manifold it doesn't explain why a second particle at rest should move towards the source mass. As such, it invites attempts at derivation from a more fundamental set of underlying assumptions as those given in this book.

Although general relativity is highly regarded for its elegance and accuracy, it has limitations: the gravitational singularities inside black holes, the ad hoc postulation of dark matter, as well as dark energy and its relation to the cosmological constant are among the current unsolved mysteries regarding gravity, all of which signal the collapse of the general theory of relativity at different scales and highlight the need for a gravitational theory that goes into the quantum realm. At distances close to the Planck length, like those near the center of a black hole, quantum fluctuations of spacetime are expected to play an important role. Quantum fluctuations are important because they contributed to the formation of galaxies. The fluctuations created slightly denser and less dense regions, and the denser regions had stronger gravitational fields. These gravitational fields pulled in matter from around them, clumping it together into gas clouds that eventually became galaxies. The imprint of these fluctuations can be seen in the cosmic microwave background radiation, the leftover radiation from the Big Bang.

The discrepancies between the predicted value for the vacuum energy and the observed values (which, depending on considerations, can be of 60 or 120 orders of magnitude) highlight the necessity for a quantum theory of gravity.

In this book we show that gravitation might be not a fundamental interaction but a byproduct of the electromagnetic interaction, precisely an electromagnetic phenomena induced by the presence of matter in the quantum vacuum (the quantum field that is present even in empty space). Which means that, matter is not just there but is in the quantum vacuum, and therefore interacts with it, causing some kind of quantum fluctuation energy, that fluctuation is gravitation. In simple terms, a body immersed in quantum fields will interact with them causing gravity to manifest.

In this work, we unify the quantum electrodynamic field theory and the Einstein gravitational field theory of general relativity into a single unified field theory of time, space and matter. The unification produces a fourth order curvature of space-time in which both the large bodies and subatomic particles play and interact. The proposed unified equation recovers the Einstein Field Equations, solves dark matter, dark energy and the cosmological constant problems. It also confirms the observed cosmological parameters by Planck collaborations in the LambdaCDM as this book directs.

Balungi Francis 2025

1

THE UNIFIED FIELD THEORY

The quantum field theory states that all of the various fundamental fields, such as the electromagnetic field, must be quantized at each and every point in space. This leaves us with a question: How do we quantize the gravitational field? Anyway do we need to quantize it or we just need to gravitise quantum mechanics? Both these questions will be answered.

For over a century, unified field theory has remained an open line of research. The term was coined by Albert Einstein, who attempted to unify his general theory of relativity with electromagnetism but who failed on grounds that he did not believe that the quantum theory was a fundamental theory for the explanation of subatomic particles of nature. He believed that it was Incomplete, a fact that has lately been taken by a most renowned physicist by the names of Sir Roger Penrose.

Although quantum mechanics may appear to be incomplete, it is one of the theories that is consistent with nature and that agrees well with experiments. This theory describes the properties and behavior of matter and radiation at their smallest scales. It is the basis for the standard model of particle physics, which builds up all the visible constituents of the universe out of collections of quarks, electrons and force-carrying particles such as photons. The discovery of the Higgs boson at CERN in Geneva is only the most recent of this theory's many truimphs.

The goal of a unified field theory has led to a great deal of progress for future theoretical physics, and progress still continues. In this book we start from where James Clerk Maxwell left us in his description of gravity and then finish his unfinished revolution. Maxwell assumed that space was made out of a material which he called ether on which particles, electromagnetic fields Interplay. This ether model was rejected later by Einstein and others but now it has come back and has been fashioned in different forms with the famous description being a quantum vacuum. Space therefore in that sense is not empty and we're yet to discover. Space or the vacuum of space is not "empty" but contains particles popping in an out of existence just like the popups on your computer screen.

According to Newton's law of Gravity, two bodies in this space will attract each other by a force of gravity that falls off as a square of distance between them. The question here is, why do the two bodies attract each other?

Maxwell's theory of electrodynamics, based on forces acting over insensible distances and on purely abstract energy-based considerations, explains why there is this attraction by saying that the energy of the system consisting of the two bodies increases when their distance increases. This idea is consistent with Maxwell's conception of the electric field in which according to the classical electromagnetic theory, the energy density of an electric field in vacuum is $\varepsilon_o E^2/2$, where E is the magnitude of the electric field at the given point and ε_o is the permittivity of the vacuum (*this result can be derived from Eqn1 below when the electric force is F=eE and the fine structure constant is* $\alpha_e = e^2/4\pi\varepsilon_0$). For the spherical field around a stationary mass point, E drops off as the square of the distance, so the energy density drops off as the fourth power, so the total integrated energy is finite.

Therefore electric field is energy in a certain state. The magnitude of energy (E) is measured by its inherent force. Energy density is the consistency of energy, which is proportional to the square of its inherent force.

Electric charge is distortion at a location. Positive distortion is balanced by negative distortion elsewhere. Therefore, the net distortion is always zero. A single isolated charge has a spherical field around it. The inherent force and consistency of this field drops off rapidly with distance, so the total energy associated with that charge is finite. For two equal and opposite charges, all lines of force start from one charge and terminate into the other. The lines do not exist when the charges are co-located. There is no charge, field or space. We are not considering the mass and material space in this situation. Suppose we separate the two oppositely charged particles by some distance. This displacement results in a net electric field in the surrounding medium. The medium or the space between the two charges now holds some energy. The energy came from the work done to separate the particles, so we see that these two oppositely charged particles exert a force of attraction on each other (through the intermediary of the dielectric medium). The further we separate the particles, the more energy we put into the field, and we approach the energy of two complete isolated fields when the particles are infinitely far apart.

However Gravitation differs from magnetism and electricity in this; that the bodies concerned are all of the same kind, instead of being of opposite signs, like magnetic poles and electrified bodies, and that the force between these bodies is an attraction and not repulsion, as is the case between like electric and magnetic bodies. To be more explicit, suppose we regard the force of gravitation as arising from the actions of a field, and suppose the presence of a gravitational field represents certain energy content. The stronger the field, the more energy it contains. Now if analyze a pair of massive particles, we find that when they are initially co-located, we have a field with twice the intensity of the field of either particle individually, and as we move the particles apart, the integral of the squared field strength (i.e., the total energy content of the field) drops, just as in the case of the electric field of two positively charged particles. Since the energy of the combined gravitational field drops as the particles are moved apart, it follows (by Maxwell's reasoning) that there is a force of repulsion, not attraction, between the particles. The force of gravity predicted by this simple energy-based reasoning is in the wrong direction. Indeed this reasoning implies that it is impossible for "like" charges to attract each other - at least if their interaction can be represented as a continuous field. Then why is gravitation an attraction?

Maxwell's failed to explain gravity simply because he didn't have the modern tools to do it. Quantum mechanics would take long to be developed and the mathematics was so young. But his ideas were phenomenal and only if he had done away with the ether description of space maybe we would have understood gravity in the same fashion as I have done in here.

Even now with the tools present, they fail because quantum mechanics is incomplete-I agree with Sir Roger Penrose on this one. The stress or energy density in Maxwell electromagnetic theory was not modified to account for the quantum vacuum fluctuations in the Quantum electrodynamics field theory. It was just instead passed on in its classical format into a newly proposed QFT, thereby coupling a classical description of matter fields in the ether to a quantum description. This is where the problems of quantization came from and we are still grinding. The description of a quantum vacuum for all fields whether gravity will have to obey the equation herein presented as new and has not been discovered anywhere:

Definition:

The universal substance is energy:

A field is energy in a certain state. The magnitude of energy (E) is measured by its inherent force. Energy density or stress in space is the consistency of energy, which is proportional to the square of its inherent force (F).

$$\rho = \frac{F^2}{8\pi\gamma\alpha_e \hbar c} = \frac{\varepsilon_0 F^2}{2\gamma e^2} \qquad (1)$$

Where $\hbar$ is the reduced Planck constant, c is the constant speed of light, $\alpha_e = e^2/4\pi\varepsilon_0\hbar c = 1/137$ is the fine structure constant and $\gamma = 2.19154$ is a dimensionless number. The square root of this value gives a value close to the reduced Hubble constant $h = \gamma^{-1/2} = 0.6755$.

Thomson proved, by strict dynamical reasoning, that the transmission of magnetic force is associated with a rotatory motion of the small parts of the medium. A theory of this kind is worked out in greater detail in Clerk Maxwell's Treatise on Electricity and Magnetism. It is there shown that, if we assume that the medium is in a state of stress, consisting of tension along the lines of force and pressure in all directions at right angles to the lines of force, the tension and the pressure being equal in numerical value and proportional to the square of the intensity of the field at the given point, the observed electrostatic and electromagnetic forces will be completely accounted for.

In the case of electromagnetic force we avail ourselves of Thomson's deduction from Faraday's discovery. We assume that the small parts of the medium are rotating about axes parallel to the lines of force. The centrifugal force due to this rotation produces the excess of pressure perpendicular to the lines of force. The explanation of electrostatic stress is less satisfactory, but there can be no doubt that a path is now open by which we may trace to the action of a medium all forces which, like the electric and magnetic forces, vary inversely as the square of the distance, and are attractive between bodies of different names, and repulsive between bodies of the same names.

After tracing to the action of the surrounding medium both the magnetic and the electric attractions and repulsions, and finding them to depend on the inverse square of the distance, we are naturally led to inquire whether the attraction of gravitation, which follows the same law of the distance, is not also traceable to the action of a surrounding medium.

Gravitation differs from magnetism and electricity in this; that the bodies concerned are all of the same kind, instead of being of opposite signs, like magnetic poles and electrified bodies, and that the force between these bodies is an attraction and not repulsion, as is the case between like electric and magnetic bodies.

The lines of gravitating force near two dense bodies are exactly of the same form as the lines of magnetic force near two poles of the same name; but whereas the poles are repelled, the bodies are attracted.

The surrounding medium has intrinsic properties in a way that it contains a virtual particle (virtual in a sense that it exists and then disappears within a short and almost unnoticeable period of time). These particles are defined as atoms of space with a weight or gravitating force which according to Newton's second law is the product of mass and acceleration due to gravity. In literal terms it is a force carried by a medium of space (Remember space is not empty). In other words these atoms of space are carriers of what is supposed to be a gravitational force, yet it is tension in space.

Let E_1 be the intrinsic energy of the field surrounding two gravitating bodies M_1 , M_2 , and let E_2 be the intrinsic energy of the field surrounding two magnetic poles , equal in numerical value to m_1 , m_2, and let F_1 be the force carried by the medium acting during the displacement δx, and F_2 the magnetic force,

$$F_1 \delta x = \delta E_1, F_2 \delta x = \delta E_2$$

now F_1 and F_2 are equal in numerical value, but of opposite signs; so that

$$\delta E_1 = -\delta E_2$$

or

$$E_1 = C - E_2$$

If F be the resultant force carried by the surrounding medium, and F_2 the resultant magnetic force at a corresponding part of the field,

$$F = -F_2$$

Hence

$$E = C - \sum \frac{1}{8\pi\gamma\alpha_e\hbar c} F^2\, dV$$

The intrinsic energy of the field of gravitation must therefore be less wherever there is a resultant force in space.

As energy is essentially positive, it is impossible for any part of space to have negative intrinsic energy. Hence those parts of space in which there is no resultant force, such as the points of equilibrium in the space between the different bodies of a system, and within the substance of each body, must have an intrinsic energy per unit of volume greater than

$$\frac{1}{8\pi\gamma\alpha_e\hbar c}F^2$$

where F is the greatest possible value of the intensity of the gravitating force (Tension) carried by the medium of space in any part of the universe. The assumption, therefore, that gravitation arises from the action of the surrounding medium in the way pointed out, leads to the conclusion that every part of this medium possesses, when undisturbed, an enormous intrinsic energy, and that the presence of dense bodies influences the medium so as to diminish this energy wherever there is a resultant stretch of the medium. The behavior of the medium at a quantum level is the cause of gravitation as we are yet to find out.

The mutual action between two bodies as pointed out is called stress. When the mutual action tends to bring the bodies nearer, or to prevent them from separating, it is called attraction. When it tends to separate the bodies, or to prevent them from approaching, it is called pressure or repulsion. The names tension and pressure are used when the action is seen to take place through a quantum vacuum.

The cause of gravitational attraction has been ruled out in various literatures to be pressure. Even Newton the Founder of the gravitational theory did not believe in the direct action of bodies at a distance. We know from his Optical Queries, that he believed that pressure in the medium was the cause of gravity. He shows that if the pressure of this medium is less in the neighborhood of dense bodies than at great distances from them, dense bodies will be drawn towards each other, and that if the diminution of pressure is inversely as the distance from the dense body the law will be that of gravitation. The next step, as he points out, is to account for this inequality of pressure in the medium ; and as he was not able to do this, he left the explanation of the cause of gravity as a problem to succeeding ages. As regards gravitation the progress made towards the solution of the problem since the time of Newton has been almost imperceptible.

The question now is:- What kind of pressure is this and where does it come from?

To answer this question we assume that the effects of gravity are associated with a rotatory motion of the small parts of the medium as in a simple harmonic oscillator. The building blocks of the medium are virtual particles with an associated mass. Therefore the small parts of the medium are atoms of space at the Planck level. It is there shown that, if we assume that the medium is in a state of stress, consisting of pressure along the lines of force and tension in all directions at right angles to the lines of force, the tension and the pressure being equal in numerical value and proportional to the square of the force of the field at the given point, the observed gravitational force will be completely accounted for.

We assume that the atoms of the medium are rotating about axes parallel to the lines of force. The centrifugal force due to this rotation produces the excess of tension perpendicular to the lines of force.

This has been dealt with in detail in a quantum field theory that; the quantum vacuum state (also called the quantum vacuum or vacuum state) is the quantum state with the lowest possible energy containing virtual particles or a zero-point field. According to quantum mechanics, the vacuum state is not truly empty but instead contains fleeting electromagnetic waves and particles that pop into and out of the quantum field. At the most basic level, the field at each point in space is a simple harmonic oscillator, and its quantization places a quantum harmonic oscillator at each point. Excitations of the field correspond to the elementary particles of particle physics.

The vacuum has, implicitly, all of the properties that a particle may have: spin, or polarization in the case of light, energy, and so on. Therefore it may be described as a single entity of matter in terms of its vacuum energy or the vacuum expectation value of the energy. The quantization of a simple harmonic oscillator therefore states that the lowest possible energy or zero-point energy that such an oscillator may have is

$$\langle E\rangle = \frac{1}{2}\hbar\sum_{n}\omega_n$$

This expectation value of energy is taken to be the intrinsic energy of the medium or the quantum vacuum.

For the case of gravity, where a coordinate system (such as normal coordinates) with coordinate radius of the order of $\Lambda^{-1/2}$at point P has been chosen (*we require this coordinate because we are aware that the subspace spanned by matter in a quantum state is for an asymptotically flat region and has no uniquely defined points in a curved spacetime. This is just a mathematical description and therefore does not mean two separate entities of space*), summing over all possible oscillators at all points in space give a finite quantity because for modes for which $\omega < c\Lambda^{1/2}$, means that the initial vacuum state $\langle E1\rangle$ is the same as the final vacuum state$\langle E3\rangle$. This then gives a constant term for the total angular frequency in Einstein space.

$$\sum_{n}\omega_n = c\Lambda^{1/2}$$

Space in Einstein gravity is equivalent to the quantum vacuum because space and the vaccum are not empty in this case. Any measurement made on space is the same as measurements done in “non’’ empty space. The quantum vacuum is not a separate entity of space. What we have done is that we have replaced empty space with non-empty space.

The intrinsic energy of the vacuum should not be confused with the intrinsic energy of the field. The former is associated with pressure while the latter is associated with tension.

Empirically the directly observed gravitational phenomena attributed to dark matter, such as the flattening of rotation curves in spiral galaxies, dark energy, the cosmological constant or the expanding expansion of the universe and the evidence from weak lensing data, occur when the vacuum energy falls below a universal value determined by the space-time curvature scale Λ:

$$\langle E \rangle = \frac{\hbar c}{2} \Lambda^{1/2}$$

Let F_3 be the quantum vacuum field force acting during the displacement δy in space

$$F_3 \delta y = \delta \langle E \rangle$$

The vacuum energy depends on the shape y of space. Each energy level E_n depends on the shape y , and so one should write $E_n(y)$ for the energy level, and ⟨E(y)⟩ for the vacuum expectation value. At this point comes an important observation: The force at point p on the shape of space is equal to the change in the vacuum energy if the shape y is perturbed a little bit, say by δy, at p. That is, one has

$$F(p) = \frac{\delta \langle \mathrm{E(y)} \rangle}{\delta \mathrm{y}} \bigg| p$$

In a limit $\delta \mathrm{y} = \Lambda^{-1/2}$, $\langle E(y) \rangle = \frac{\hbar c}{2} \Lambda^{1/2}$

$$F_3 = \frac{\hbar c}{2}\Lambda$$

The fact that the resulting force F_3 is very small is proof that we are dealing with a universe described by an effective local quantum field theory down to the Planck scale. The space-time at the Planck level admits great fluctuations of the metric and therefore leads to the possibility of the creation of many virtual particles.

The fact that there is no area that can be measured below a minimum amount indicates that there is as sort of minimal size of physical space, at the Planck scale. This discreteness of physical space at the Planck length has long been expected in Quantum Gravity. Space geometry is therefore quantized in the same manner in which the energy of a harmonic oscillator is quantized. This then implies a maximal angular frequency for particle creation by curvature due to quantum fluctuations of a quantum field in a Planck scale universe of;

$$\sum_n \omega_n = c\mathrm{B}^{1/2}$$

Where, $B^{-1/2}$ imposes a general minimal bound on the radius of curvature in a Planck scale universe when a coordinate radius $l = B^{-1/2}$ has been chosen for any point say p in the physical space at distances close to the Planck length.

Notice that in the context of LQG this discreteness is not imposed or postulated. Rather, it is a direct consequence of a straightforward quantization of GR. Then the total surface spherical area of space is calculated to be

$$A = \frac{4\pi \mathrm{B}^{-3/2}}{c} \sum_n \omega_n$$

Hence

$$A = 4\pi \mathrm{B}^{-1}$$

But since pressure is force per unit surface area then the intrinsic pressure of the medium is given as,

$$P = \frac{F_3}{A}$$

Or

$$P = \frac{\hbar c}{8\pi} B\Lambda$$

This pressure is equivalent to the local vacuum energy density of particles created by a gravitational field The vacuum pressure is small and constant irrespective of how big or small the curvature is. However, as was shown above, the energy density of the created particles is small locally irrespective of the curvature which created them. Although the radius of curvature $\Lambda^{-1/2}$ is large compared to the length $l = \mathrm{B}^{-1/2}$. The product of $\hbar$ and Λ cancles out the large value of B to produce a small vacuum energy density. One would therefore expect that the scheme of calculating the effect of quantum vacuum fluctuations occurring at the Planck scale (in regions where the curvature B is high) on our observable universe described by a cosmological constant in curved space time is a good approximation for a quantum theory of gravity.

From the classical singularity theorems , one would expect such high curvatures to occur in collapsing stars and, in the past, at the beginning of the present expansion phase of the universe. One therefore would expect the regions of high curvature to be hidden from us by an event horizon. Thus, as far as we are concerned, this quantum geometry of matter treatment should be valid up to the first $10^{-43}s$ of the universe.

It is here shown that, if we assume that the medium is in a state of stress, consisting of pressure along the lines of force and tension in all directions at right angles to the lines of force, the tension and the pressure being equal in numerical value and proportional to the square of the intensity of the gravitating force carried by the medium in any part of the universe of the field at the given point, the observed gravitational effects will be completely accounted as:

$$\frac{C}{dV} - \frac{E}{dV} = \sum \frac{1}{8\pi\gamma\alpha_e\hbar c} F^2$$

Where

$$P = \frac{F^2}{8\pi\gamma\alpha_e\hbar c}$$

$$\frac{\hbar c}{8\pi} B\Lambda = \frac{F^2}{8\pi\gamma\alpha_e\hbar c}$$

$$B\Lambda = \frac{F^2}{\gamma\alpha_e(\hbar c)^2} \qquad (2)$$

We have therefore explained gravity in terms of quantum mechanics. The unification of the quantum electrodynamic field theory and the gravitational field theory imposes a fourth order curvature $\phi = B\Lambda$ (measured in units of $length^{-4}\ or\ m^{-4}$) and is related to the energy density ρ of matter fields as:

$$\phi = \frac{8\pi}{\hbar c}\rho$$

From Eqn2 we have the Resultant gravitational force carried by the medium as,

$$F = \hbar c(\gamma\alpha_e B\Lambda)^{1/2}$$

1.1.1. Calculation of the amount of stress which would be required to account for the actual effects of gravity at the surface of the earth.

It is easy to calculate the amount of the stress which would be required to account for the actual effects of gravity at the surface of the earth. As we postulated before and explained from a quantum point of view, the building blocks of space and any medium surrounding like bodies is termed as atoms of space with a measurable mass. The physical cause of such a state of stress is from quantum vacuum fluctuation that leads to the creation of particles or matter or mass by curvature or simply the local energy density of particles created by a gravitational field. At high curvature B, the particle created is a Planck-mass-sized black hole remnant or a virtual black hole, which is a hypothetical micro black hole that exists temporarily as a result of a quantum fluctuation of spacetime.

Such Planck-mass black holes may in effect be stable objects if the quantized gaps between their allowed energy levels bar them from emitting Hawking particles or absorbing energy gravitationally like a classical black hole. In such case, they would be weakly interacting massive particles; which in this case could be particle candidates for dark matter and dark energy.

At the surface of the earth, the acceleration due to gravity is from Newton's theory and is given as $g =$

$9.807m/s^2$. The mass of the atom of the medium which in this case are Planck mass particles is given also by $m_p = \left(\frac{\hbar c}{G}\right)^{1/2} = 2.1765 \times 10^{-8}kg$. Then the force or weight of the medium surrounding the earth is given by

$$F = m_p g = \left(\frac{\hbar c}{G}\right)^{1/2} g = 2.134494 \times 10^{-7} N$$

But since from Eqn1 tension which is in all horizontal directions is given by

$$\rho = \frac{F^2}{8\pi\gamma\alpha_e \hbar c} = \frac{\left(m_p g\right)^2}{8\pi\gamma\alpha_e \hbar c}$$

Then

$$\rho = 3.5805 \times 10^{12} N/m^2$$

According to the theory herein presented, the above given tension must be equal in the same numerical value to the pressure in the vertical direction. When this balance is achieved then the effects of gravity will manifest.

$$P = \frac{\hbar c}{8\pi} B\Lambda$$

The curvature of space according to Einstein's space-time curvature calculated at the Earth's surface for our solar system is $\Lambda = 7.423 \times 10^{-31} m^{-2}$. The curvature required to produce Planck mass particles at a radius of curvature close to the Planck length is, $\Lambda = 0.383 \times 10^{70} m^{-2}$. These values give a pressure $P = 3.5805 \times$

$10^{12} N/m^2$ in a vertical derection of the same numerical value as tension

The obtained value is in agreement with Maxwell's value of pressure of 37,000 tons weight on the square inch in a vertical direction. Though Maxwell's value is less by -3N/m^2 (***See Attraction by James Clerk Maxwell in Encyclopedia Britannica, Ninth Edition, Volume III***). Maxwell calculated pressure by assuming a weight of the earth per unit surface area of the earth as, $P = \frac{m_E g}{4\pi {R_E}^2}$.

1.1.2.Einstein Field Equation from a Unified Equation

In the case of gravitational force we assume that the small parts of the medium are rotating about axes parallel to the lines of force. The centrifugal force due to this rotation produces the excess of tension perpendicular to the lines of force. To account for such a force by means of stress in an intervening medium, we must suppose that there is a pressure in the direction of the lines of force, combined with a tension in all directions at right angles to the lines of force. Such a state of stress would, no doubt, account for the observed effects of gravitation. The physical cause of such a state of stress is from quantum vacuum fluctuation that leads to the creation of particles or matter or mass by curvature or simply the local energy density of particles created by a gravitational field. At high curvature B, the particle created is a Planck-mass-sized black hole remnant or a virtual black hole, which is a hypothetical micro black hole that exists temporarily as a result of a quantum fluctuation of spacetime. The basic idea here is that at the Planck level space-time admits great fluctuations of the metric and therefore leads to the possibility of the creation of many (virtual) particles. That is to say, of the many particles created, the curvature B also creates micro-black holes. Such Planck-mass black holes may in effect be stable objects if the quantized gaps between their allowed energy levels bar them from emitting Hawking particles or absorbing energy gravitationally like a classical black hole. In such case, they would be weakly interacting massive particles; which in this case could explain dark matter and dark energy.

This situation only happens at the Planck scale at lengths close to the Planck length of $l = \mathrm{B}^{-1/2} = \sqrt{\frac{\hbar G}{c^3}}$

When all this has been put into consideration we recover the Einstein Field Equation in this way,

$$\Lambda = \frac{8\pi B^{-1}}{\hbar c}\rho$$

This implies that high curvatures in the virtual dimension (Planck scale universe) produce many particles leading to great vacuum fluctuations. The pressure is in the direction of the lines of force, combined with a tension in all directions at right angles to the lines of force. Such a state of stress would, no doubt, account for the observed effects of gravitation as was depicted by Albert Einstein

$$\Lambda = \frac{8\pi G}{c^4}\rho$$

Where Λ denotes the total constant curvature of space time (or the cosmological constant) , G is the Gravitational constant and is the ρ energy density. The cosmological constant has the same effect as an intrinsic energy density of the vacuum, ρ (and an associated pressure). In this context, it is commonly moved to the right-hand side of the equation using $\Lambda = \kappa\rho$. It is common to quote values of energy density directly, though still using the name "cosmological constant". The dimension of Λ is generally understood as length^{-2}.

The cosmological constant Λ appears in the Einstein field equations in the form

$$R_{\mu\nu} - \frac{1}{2}Rg_{\mu\nu} + \Lambda g_{\mu\nu} = \kappa T_{\mu\nu}$$

where the Ricci tensor $R_{\mu\nu}$, Ricci scalar R and the metric tensor $g_{\mu\nu}$ describe the structure of spacetime, the stress-energy tensor $T_{\mu\nu}$ describes the energy density, momentum density and stress at that point in spacetime, and $\kappa = 8\pi G/c4$. The gravitational constant G and the speed of light c are universal constants. When Λ is zero, this reduces to the field equation of general relativity usually used in the 20th century. When $T\mu\nu$ is zero, the field equation describes empty space (a vacuum).

1.2. The Cosmological Constant from the Unified equation

According to quantum field theory (QFT), empty space is not empty in a sense that is defined by the vacuum state, which is composed of a collection of quantum fields. All these quantum fields exhibit fluctuations of virtual subatomic particles popping in and out of existence in their ground state (lowest energy density) arising from the zero-point energy existing everywhere in space. In this section we prove that these zero-point fluctuations contribute to the cosmological constant Λ because as in the previous sections we believe that the vacuum holds the key to a full understanding of gravity and nature

A common assumption is that the quantum vacuum (or vacuum energy density fluctuations ρ) is equivalent to the cosmological constant Λ as modeled in (Eqn2)

$$\Lambda = \frac{F^2}{\gamma \alpha_e (\hbar c)^2 B}$$

Therefore for the universe described by an effective local quantum field theory down to the Planck scale, we will produce a cosmological constant of the order of ${m_{pl}}^2$ in (1 reduced Planck units) as described below

Consider a Planck-mass-sized black hole remnant or a virtual black hole, which is a hypothetical micro black hole that exists temporarily as a result of a quantum fluctuation of spacetime. The basic idea here is that at the Planck level space-time admits great fluctuations of the metric and therefore leads to the possibility of the creation of many (virtual) particles. That is to say, of the many particles created, the curvature B also creates micro-black holes. Such Planck-mass black holes may in effect be stable objects if the quantized gaps between their allowed energy levels bar them from emitting Hawking particles or absorbing energy gravitationally like a classical black hole. In such case, they would be weakly interacting massive particles; which in this case could explain dark matter and dark energy.

The tension of the medium is proportional to the square of the force carried by the medium or field at the given point. This force carries a Planck mass particle and can be termed as the weight the medium in the gravitational field, a_o. Then by Newton's second law:

$$F = m_{pl} a_o$$

Where $m_{pl} = \sqrt{\frac{\hbar c}{G}}$, is the Planck mass and $a_o = 1.2 \times 10^{-10} m/s^2$ is a constant MOND acceleration (we have assumed that these particles are weakly interacting massive particle and are therefore candidates for dark matter and dark energy). Then on substitution into (Eq2) we get:

$$\Lambda = \frac{{m_{pl}}^2 {a_o}^2}{(\hbar c)^2 \gamma \alpha_e B} \propto {m_{pl}}^2$$

And on arranging and cancelling like terms we get:

$$\Lambda = \frac{{a_o}^2}{\gamma \alpha_e c^4} \qquad (3)$$

$$\Lambda = 1.11 \times 10^{-52} m^{-2}$$

This value is also in good agreement with the recently observed **Planck (2018)** collaboration cosmological constant values and Planck units for ΩΛ = 0.6889±0.0056 and the Hubble constant $H_o = 67.66 \pm 0.42\ (km/s)/Mpc =\ H_o = 2.1928 \times 10^{-18} s^{-1}$

1.2.1. The vacuum energy density of Dark energy

Our equation (Eqn1) also implies a vacuum energy density of dark energy for Planck sized particles of space in a gravitational field of a constant MOND acceleration with a force or weight $F = m_{pl} a_o$ where, m_{pl} - the Planck mass. We then have the energy density of dark energy as;

$$\rho = \frac{m_{pl}{}^2 a_o{}^2}{8\pi\gamma\alpha_e\hbar c} = \frac{a_o{}^2}{8\pi G\gamma\alpha_e}$$

$$\rho = 5.371 \times 10^{-10} \frac{J}{m^3}$$

This value is in good agreement with the observed Planck (2015b) collaboration cosmological constant values. The above equation is similar but differs in constants from the Verlinde's energy density result as outlined in his popular paper "Emergent Gravity and the Dark Universe".

1.4. The relationship between the Hubble Constant and the MOND acceleration

In this section we want to express the MOND acceleration in terms of the Hubble constant in a bid to prove that this acceleration is not only related to Dark matter but it is also an inherent property of space time.

In the standard model of Cosmology known as the ΛCDM model, the cosmological constant has the value of;

$$\Lambda = 3\left(\frac{\mathrm{H}_o}{c}\right)^2 \Omega_\Lambda = 1.11 \times 10^{-52} m^{-2}$$

Where $\mathrm{H}_o = 2.1928 \times 10^{-18} s^{-1}$ is the Hubble constant and $\Omega_\Lambda = 0.6889$ is the omega Sub lambda or Density Parameter), these are recently observed Planck **(2018)** collaboration cosmological constant values and Planck units for ΩΛ = 0.6889±0.0056 and the Hubble constant $\mathrm{H}_o = 67.66 \pm 0.42\ (km/s)/Mpc = \quad \mathrm{H}_o = 2.1928 \times 10^{-18} s^{-1}$

The above given value is in agreement with our previously derived value in Eqn3 given as;

$$\Lambda = \frac{{a_o}^2}{\gamma \alpha_e c^4}$$

Equating the above two equations, we can create a relationship between the MOND acceleration and the Hubble constant as,

$$a_o = \mathrm{H}_o c \sqrt{3\gamma \alpha_e \Omega_\Lambda} \quad (5)$$

$$a_o = 1.2 \times 10^{-10} m/s^2$$

This value is in good agreement with experiment.

1.5. Evidence of Dirac large numbers hypothesis

In this section we deduce that the ratio of the maximum acceleration in Quantum gravity to the MOND acceleration is equal to the ratio of the Planck mass to the minimum mass of a particle and also equal to the ratio of the Planck mass to the mass of the Hubble sphere. The relationship is deduced by equating the Equation for the Cosmological constant Lambda CDM to our unified Eqn3 as

$$\Lambda = 3\left(\frac{\mathrm{H}_o}{c}\right)^2 \Omega_\Lambda = \frac{1}{(\hbar c)^2}\left(\frac{1}{\gamma \alpha_e}\right)\left(m_{pl} a_o l_p\right)^2$$

Taking square roots on both sides of the equation we have on arranging, the ratio of the ratio of the maximum acceleration in Quantum gravity a_{max} to the MOND accelerationa_o is equal to the ratio of the Planck mass m_{pl} to the minimum mass of a particle in Quantum cosmology (the Hubble scale mass) m_s and also equal to the ratio of the Planck mass m_{pl} to the mass of the Hubble sphere M_H equal to a very large number denoted N.

$$\frac{a_{max}}{a_o} = \frac{m_{pl}}{m_s} = \sqrt{\frac{c^5}{\hbar G \mathrm{H}_o{}^2}} = \frac{m_{pl}}{M_H} = 8.47 \times 10^{60}$$

$$N = 8.47 \times 10^{60}$$

Where;

$a_{max} = \frac{c^2}{l_p}\sqrt{3\gamma\alpha_e\Omega_\Lambda} = 1.0227 \times 10^{51} m/s^2$, is the maximum acceleration in Quantum Gravity.

$m_s = \frac{\mathrm{H_o}\hbar}{c^2} = 2.57 \times 10^{-69} kg$, is the minimum mass possible in Quantum Cosmology (the Hubble scale mass). This exceptionally small mass coincides with the minimal measurable gravitational self-energy of a particle which is accepted as minimum quantum of energy$E_{min} = \mathrm{H_o}\hbar = 10^{-33}\mathrm{eV}$. This energy takes substantial place in the estimations of total information and entropy of the observable universe . Thus, the mass m_s seems close to the graviton mass obtained by different methods. The mass m_s is several orders of magnitude smaller than the upper limit of graviton mass, obtained by astrophysical constraints. Therefore, this value doesn't contradict of astrophysical observations.

$M_H = \frac{c^3}{H_o G} = 1.846 \times 10^{53} kg$, is the critical mass of the observable universe.

These ratios are very important because they connect cosmological parameters (mass of the Hubble sphere, Hubble scale mass and MOND acceleration and the respective fundamental microscopic properties of matter (Planck mass, Planck length and Planck maximum acceleration in Quantum gravity). In recent quantum gravity models, the Planck units imply quantization of spacetime at extremely short range. Thus, the ratios represent connection between cosmological parameters and quantum properties of spacetime.

Finally the ratio of M_H to m_s gives us a complete and final Equation connecting the micro world to the macro world in this way

$$\frac{M_H}{m_s} = \frac{c^5}{\hbar G {H_o}^2} = \frac{{m_{pl}}^2}{{M_H}^2} = N^2 = 7.174 \times 10^{121}$$

The above calculated large number is also the ratio of Planck density and recent critical density of the universe.

Which when arranged gives an empirical relation

$${M_H}^3 = {m_{pl}}^2 m_s$$

Connecting the large scale M_H to the small scale m_{pl} and m_s we have

$$M_H = \left({m_{pl}}^2 m_s\right)^{1/3}$$

Thus, a precise formulation and evidence of Dirac Large number hypothesis has been found connecting the micro world and the macro world.

Since the most of parameters entering equation the ratios are characterized by the maximum or minimum possible values in the universe, it can be roughly claimed that the Dirac LNH shows that the largest and the smallest in the universe are related with a number of the order of $N = 8.47 \times 10^{60}$

1.5.1. Calculating the Cosmological constant using the Hubble scale mass and the maximum acceleration in Quantum Gravity

Now assume that the graviton or a Hubble scale mass particle $m_s = \frac{\mathrm{H_o}\hbar}{\mathrm{c}^2} = 2.57 \times 10^{-69} kg$, accelerates with an associated maximum constant acceleration ($a_{max} = \frac{c^2}{l_p}\sqrt{3\gamma\alpha_e\Omega_\Lambda} = 1.0227 \times 10^{51} m/s^2$) when a force (F) is applied to it according to Newton's second law, $F = m_s a_{max}$. Then the value of the cosmological constant according to the unified equation is

$$\Lambda = \frac{1}{(\hbar c)^2}\left(\frac{1}{\gamma\alpha_e}\right)\left(m_s a_{max} l_p\right)^2$$

$$\Lambda = 3\left(\frac{\mathrm{H}_o}{c}\right)^2 \Omega_\Lambda$$

$$\Lambda = 1.11 \times 10^{-52} m^{-2}$$

This value is also in good agreement with the recently observed **Planck (2018)** collaboration cosmological constant values and Planck units for ΩΛ = 0.6889±0.0056 and the Hubble constant $\mathrm{H}_o = 67.66 \pm 0.42\ (km/s)/Mpc = \ \mathrm{H}_o = 2.1928 \times 10^{-18} s^{-1}$

1.5.2. Calculation of the vacuum energy density of Dark energy using the Hubble scale mass and the maximum acceleration in Quantum Gravity

Our Unified equation also implies a vacuum energy density of dark energy for gravitons or Hubble scale mass particles accelerating with a constant maximum acceleration with a force $F = m_s a_{max}$. We then have the energy density of dark energy as;

$$\rho = \frac{{m_s}^2 {a_{max}}^2}{8\pi\gamma\alpha_e \hbar c} = \frac{3{H_o}^2 c^2}{8\pi G}\Omega_\Lambda$$

$$\rho = 5.371 \times 10^{-10} \frac{J}{m^3}$$

It is therefore theorized that, gravitons or Hubble scale mass particles accelerate with a maximum acceleration of $a_{max} = 1.0227 \times 10^{51} m/s^2$ in Planck sized space whereas Planck mass sized particles accelerate with a minimum MOND acceleration of $a_o = 1.2 \times 10^{-10} m/s^2$ in Planck sized space. Both scenarios lead to a constant cosmological constant.

1.5.3. The Definition of the MOND acceleration constant

From our relationship of the Dirac Large numbers as was deduced above we clearly notice that the MOND acceleration is defined as a gravitational field in which a Planck mass particle feels a microscopic Force of $F = m_s a_{max} = 2.628 \times 10^{-18} N$. In simple terms, by Newton's second law the MOND acceleration is the force per unit mass as,

$$a_o = \frac{m_s a_{max}}{m_{pl}}$$

In conclusion here; Both the Planck mass particle and the Hubble Scale mass particle feel the same force though accelerating in different gravitational fields.

1.6. The relationship between the Cosmological Constant and the MOND acceleration

The Hubble constant in the Lambda CDM model above is given as

$$H_o = c\sqrt{\frac{\Lambda}{3\Omega_\Lambda}}$$

Substituting this value into our MOND acceleration (Eqn5) we then get a relation that has baffled scientists for years as

$$a_o = c^2\sqrt{\gamma\alpha_e\Lambda} = c^2\sqrt{\frac{\Lambda}{62.5131}}$$

This value differs widely from the popularized MOND cosmological relation of $a_o \sim \sqrt{\frac{\Lambda}{3}}$. In conclusion According to the relation derived above, this popularized relation is wrong.

2

MASS OF ORDINARY MATTER IN THE UNIVERSE EQUATION

The total mass of ordinary matter in the universe is estimated to be about 1.5 x 10^53 kilograms. This is based on the critical density and the diameter of the observable universe. The values for the critical density, the volume and diameter of the universe can not be measured accurately and they are bound to change with time. This means that they are not constants.

In this section we want to derive the mass of the observable universe without using its volume and diameter. The value we shall get will be in agreement with the Lambda CDM values. To do this, we assume a homogeneous, isotropic, flat Universe and using the fundamental constants for Gravitation (G), Cosmological constant (Λ), and the MOND acceleration (a_o) constant , dimensional analysis gives an estimated mass of

$$M_U = \frac{a_o}{\beta \Lambda G} = \frac{a_o}{\gamma \Omega_b \Lambda G} \qquad (6)$$

Where $\beta = 0.1081$ is a number determined by experiment. This value is related to the Baryon Density and the *pergravitivity of the vacuum.* $\beta = 0.1081 \approx$

$\gamma\Omega_b$ *(Where* Ω_b *is the Baryon Density Parameter* $\Omega_b \approx 0.04933$ *and* $\gamma = 2.19154$*)*

The equation for the total mass of ordinary matter in universe contains a term for the MOND acceleration Eqn5. We would like to eliminate this acceleration and thereby create an equation for the mass of matter; we do this by substituting Eqn5 into Eqn6 above

$$M_U = \frac{H_o c}{\Omega_b \Lambda G}\sqrt{\frac{3\Omega_\Lambda \alpha_e}{\gamma}}$$

But since in the ΛCDM model, the cosmological constant has the value of $\Lambda = 3\left(\frac{\mathrm{H}_o}{c}\right)^2 \Omega_\Lambda$, then

$$M_U = \frac{c^3}{H_o G}\sqrt{\frac{\alpha_e}{3\gamma{\Omega_b}^2\Omega_\Lambda}} = 1.5 \times 10^{53}\mathrm{kg} \quad (7)$$

This value has been calculated using the cosmological parameters , $\gamma = 0.4563$, $\Omega_b = 0.04933$, $\Omega_\Lambda = 0.6889$ and $\mathrm{H_o} = 2.1928 \times 10^{-18}s^{-1}$ (Planck(2018)). The mass of the observable universe hereby derived is independent of the critical density, volume and diameter of the Universe yet it corresponds to the mass that was calculated using the Planck units in Lambda CDM of 4.8% of the total critical density($0.85 \times 10^{-26}\mathrm{kg/m^3}$) at $\mathrm{H_o} = 2.176 \times 10^{-18}s^{-1}$ or$4.08 \times 10^{-28}\, kg/\mathrm{m}^3$ and using the commoving distance (radius) of about 46.6 billion light-years, they got a volume equals $3.58 \times 10^{80}\mathrm{m}^3$ and the mass of ordinary matter equals density ($4.08 \times 10^{-28}\, kg/\mathrm{m}^3$) times volume ($3.58 \times 10^{80}\mathrm{m}^3$) or $1.46 \times 10^{-53}\, kg$.

2.1. Derivation of the Fine structure constant

We have seen the appearance of the fine structure constant in our equations, the question is, does this constant in any way related to the cosmological parameters?

In this section we derive the fine structure constant by rewriting Eqn7 above in the form

$$M_U = M_H \sqrt{\frac{\alpha_e}{3\gamma{\Omega_b}^2\Omega_\Lambda}}$$

Where M_H is the mass of the Hubble sphere $M_H = \frac{c^3}{H_o G}$. The Hubble sphere is a spherical region of the universe surrounding an observer beyond which objects recede from that observer at a rate greater than the speed of light.

This implies that the mass of the observed universe is the mass of the Hubble sphere times the square root of some constant of nature related to one of the cosmological parameters we are yet to find out.

The ratio of the observable mass of the universe to the mass of the Hubble sphere is equal to a number as

$$\frac{M_U}{M_H} = \sqrt{\frac{\alpha_e}{3\gamma{\Omega_b}^2\Omega_\Lambda}} = 0.81387 \qquad (8)$$

The number given above has been purely derived from the known constants given in the theory developed above and nothing new has been added.

By observation, it follows that this number is the same as; $\sigma_8 = 0.8133$ (The present root-mean- square matter fluctuation, averaged over a sphere of radius 8h-1Mpc) and $a^{1/2} = 0.8136$ where a is calculated from; the transition from decelerating to accelerating expansion (the second derivative crossing zero) occurred when

$$a = \left(\frac{\Omega_m}{2\Omega_\Lambda}\right)^{1/3} \sim 0.662$$

The square root of a gives the value $a^{1/2} = 0.8136.$

Also, the S_8 parameter in the ΛCDM model quantifies the amplitude of matter fluctuations in the late universe and is defined as

$$S_8 = \sigma_8 \sqrt{\frac{\Omega_m}{0.3}}$$

Where the constant $\sigma_8 = 0.8133$ (The present root-mean-square matter fluctuation, averaged over a sphere of radius 8h-1Mpc).

Therefore connecting the above density parameters into our mass ratio formula we have

$$\frac{\boldsymbol{M_U}}{\boldsymbol{M_H}} = \sqrt{\frac{\boldsymbol{\alpha_e}}{\boldsymbol{3\gamma\Omega_b{}^2\Omega_\Lambda}}} = \left(\frac{\boldsymbol{\Omega_m}}{\boldsymbol{2\Omega_\Lambda}}\right)^{1/6} = \boldsymbol{S_8}\sqrt{\frac{\boldsymbol{0.3}}{\boldsymbol{\Omega_m}}} = \boldsymbol{\sigma_8} = \boldsymbol{0.81387}$$

From the above given equality we can get the value of the fine structure constant purely fashioned in the known density parameters as

$$\alpha_e = 3(\gamma\Omega_b{}^2)\left(\frac{\Omega_m\Omega_\Lambda{}^2}{2}\right)^{\frac{1}{3}} = \frac{0.9\gamma\Omega_b{}^2\Omega_\Lambda S_8{}^2}{\Omega_m}$$

$$= 3(\gamma \Omega_b{}^2)\Omega_\Lambda \sigma_8{}^2$$

(8)

Where, $\Omega_\Lambda = 0.6889$, $\sigma_8 = 0.8133$, $\Omega_b = 0.04933$, $\gamma = 2.19154$

The electromagnetic coupling constant is therefore given as

$$\boldsymbol{\alpha_e = 3\gamma\Omega_\Lambda(\Omega_b\sigma_8)^2 = 0.007299 \sim 1/137}$$

The above given value is in agreement with experiment. This result is proof that our model is consistent with nature and it also proves that the cosmological parameters herein used are true constants of nature and the Lambda CDM model is valid at the 2018 Planck units.

2.2. The Final Equation; The one Inch Equation of the Universe (A connection between the cosmological constant , the Hubble constant, the MOND acceleration constant, the fine structure constant, the gravitational constant and the density parameters)

Equating Eqn6 to Eqn7, we get a connection between the cosmological constant , the Hubble constant, the MOND acceleration constant, the fine structure constant, the density parameters and the Gravitational constant as,

$$G = \frac{a_o}{\Lambda M_H}\sqrt{\frac{3\Omega_\Lambda}{\gamma\alpha_e}}$$

But Since $M_H = \frac{c^3}{H_o G}$, then the cosmological constant is related to all the constants of nature by

$$\Lambda = \frac{a_o H_o}{c^3}\sqrt{\frac{3\Omega_\Lambda}{\gamma\alpha_e}} \qquad (11)$$

This work has been performed during the past 4 years and I have benefited from discussions and received encouragement from many colleagues. I like to thank various astronomers and cosmologists whose knowledge benefitted this work and whose support I have appreciated greatly.
The authors declare that they have no known competing financial interests or personal relationships

that could have appeared to influence the work reported in this paper.

ADDITIONAL READINGS

Balungi Francis, (2010) "A hypothetical investigation into the realm of the microscopic and macroscopic universes beyond the standard model" general physics arXiv:1002.2287v1 [physics.gen-ph]

Hawking, Stephen (1975). "Particle Creation by Black Holes". Commun. Math. Phys. 43 (3): 199-220. Bibcode:1975CMaPh..43..199H.

Hawking, S. W. (1974). "Black hole explosions?". Nature.248(5443):30-31.
Bibcode:1974Natur.248...30H.doi:10.1038/248030a0.

Carlo Rovelli (2003) "Quantum Gravity" Draft of the Book Pdf
Some few texts used are from Wikipedia https://creativecommons.org/licenses/by-sa/3.0/
D. N. Page, Phys. Rev. D 13, 198 (1976).

C. Gao and Y.Lu, Pulsations of a black hole in LQG (2012) arXiv:1706.08009v3

A.H. Chamseddine and V.Mukhanov, Non singular black hole (2016) arXiv 1612.05861v1

M.Bojowald and G.M.Paily, A no-singularity scenario in LQG (2012) arXiv: 1206.5765v1

P.Singh, class.Quant.Grav,26,125005(2009), arXiv:0901.2750

P.Singh and F.Vidotto, Phys.Rev, D83,064027(2011) arXiv:1012.1307

C.Rovelli and F.Vidotto, Phy. Rev,111(9) 091303(2013) arXiv:1307.3228v2

M.Bojowald, Initial conditions for a universe, Gravity Research Foundation (2003)

A.Ashtekar, Singularity Resolution in Loop Quantum Cosmology (2008) arXiv:0812.4703v1

J.Brunneumann and T.Thiemann, On singularity avoidance in Loop Quantum Gravity (2005) arXiv:0505032v1

L.Modesto, Disappearence of the Black hole singularity in Quantum gravity (2004) arXiv:0407097v2

Mikhailov, A.A. (1959).Mon. Not. Roy. Astron. Soc.,119, 593.

P. Merat etal.(1974). Astron & Astrophys 32, 471-475

Trempler, R.J. (1956).Helv. Phys. Acta, Suppl.,IV, 106.

Trempler, R.J. (1932). " The deflection of light in the sun's gravitational field "Astronomical society of the pacific 167

Einstein, A. (1916).Ann. d. Phys.,49, 769; (1923).The Principle of Relativity, (translators Perret, W. and Jeffery, G.B.), (Dover Publications, Inc., New York), pp. 109-164.

Von Klüber, H. (1960). InVistas in Astronomy, Vol. 3, pp. 47-77.

K. Hentschel (1992). Erwin Finlay Freundlich and testing Einstein theory of relativity, Communicated by J.D. North

Muhleman, D.O., Ekers, R.D. and Fomalont, E.B. (1970).Phys. Rev. Lett.,24, 1377

Mikhailov, A.A. (1956).Astron. Zh.,33, 912.

Dyson, F.W., Eddington, A.S. and Davidson, C. (1920).Phil. Trans. Roy. sog., A220, 291

Chant, C.A. and Young, R.K. (1924).Publ. Dom. Astron. Obs.,2, 275.

Campbell, W.W. and Trumbler, R.J. (1923).Lick Obs. Bull.,11, 41.

Freundlich, E.F., von Klüber, H. and von Brunn, A. (1931).Abhandl. Preuss. Akad. Wiss. Berlin, Phys. Math. Kl., No.l;Z. Astrophys.,3, 171

Mikhailov, A.A. (1949).Expeditions to Observe the Total Solar Eclipse of 21 September, 1941, (report), (ed. Fesenkov, V.G.), (Publications of the Academy of Sciences, U.S.S.R.), pp. 337-351.

S.P. Martin, in Perspectives on Supersymmetry , edited by G.L. Kane (World Scientific, Singapore, 1998) pp. 1-98; and a longer archive version in hep-ph/9709356; I.J.R. Aitchison, hep-ph/0505105.

Mara Beller, Quantum Dialogue: The Making of a Revolution. University of Chicago Press, Chicago, 2001.

Morrison, Philp: "The Neutrino, scientific American, Vol 194,no.1 (1956),pp.58-68.

R. Haag, J. T. Lopuszanski and M. Sohnius, Nucl. Phys. B88, 257 (1975) S.R. Coleman and J. Mandula, Phys.Rev. 159 (1967) 1251.

H.P. Nilles, Phys. Reports 110, 1 (1984).

P. Nath, R. Arnowitt, and A.H. Chamseddine, Applied N = 1 Supergravity (World Scientific, Singapore, 1984).

S.P. Martin, in Perspectives on Supersymmetry , edited by G.L. Kane (World Scientific, Singapore, 1998) pp. 1-98; and a longer archive version in hep-ph/9709356; I.J.R. Aitchison, hep-ph/0505105.

S. Weinberg, The Quantum Theory of Fields, VolumeIII: Supersymmetry (Cambridge University Press, Cambridge,UK, 2000).

E. Witten, Nucl. Phys. B188, 513 (1981).

S. Dimopoulos and H. Georgi, Nucl. Phys. B193, 150(1981).

N. Sakai, Z. Phys. C11, 153 (1981);R.K. Kaul, Phys. Lett. 109B, 19 (1982).

L. Susskind, Phys. Reports 104, 181 (1984).
L. Girardello and M. Grisaru, Nucl. Phys. B194, 65(1982); L.J. Hall and L. Randall,

Phys. Rev. Lett. 65, 2939(1990);I. Jack and D.R.T. Jones, Phys. Lett. B457, 101 (1999).

For a review, see N. Polonsky, Supersymmetry: Structureand phenomena. Extensions of the standard model, Lect.Notes Phys. M68, 1 (2001).

G. Bertone, D. Hooper and J. Silk, Phys. Reports 405, 279 (2005).

G. Jungman, M. Kamionkowski, and K. Griest, Phys. Reports 267, 195 (1996).

V. Agrawal, S.M. Barr, J.F. Donoghue and D. Seckel,Phys. Rev. D57, 5480 (1998).

N. Arkani-Hamed and S. Dimopoulos, JHEP 0506, 073(2005); G.F. Giudice and A. Romanino, Nucl. Phys. B699, 65(2004) [erratum: B706, 65 (2005)]. July 27, 2006 11:28

en.wikipedia.org/wiki/Supersymmetry - 52k - Cached - Similar pages
en.wikipedia.org/wiki/Grand_unification_theory - 39k - Cached - Similar pages

In cosmology, the Planck epoch (or Planck era), named after Max Planck, is the earliest period of time in the history of the universe, en.wikipedia.org/wiki/**Planck_epoch** - 23k - Cached - Similar pages

L. Shapiro and J. Sol`a, Phys. Lett. B 530, 10 (2002);

E. V.Gorbar and I. L. Shapiro, JHEP 02, 021 (2003); A. M. Pelinson,

L. Shapiro, and F. I. Takakura, Nucl. Phys. B 648, 417 (2003).

Carter, B.: Axisymmetric black hole has only two degrees of freedom. Phys. Rev. Lett. 26, 331-333(1971)

Penrose, R.: Gravitational collapse: the role of general relativity. Riv. Nuovo Cim. 1, 252-276 (1969)

Christodoulou, D.: Reversible and irreversible transformations in black hole physics. Phys. Rev. Lett. 25, 1596-1597 (1970)

Christodoulou, D., Ruffini, R.: Reversible transformations of a charged black hole. Phys. Rev. D4, 3552-3555 (1971)

Hawking, S.: Particle creation by black holes. Commun. Math. Phys. 43, 199-220 (1975)

Wald, R.M.: General Relativity. The University of Chicago Press, Chicago (1984)

Hawking, S.W.: Black hole explosions. Nature 248, 30-31 (1974)

Hawking, S., Ellis, G.: The large scale structure of space-time. Cambridge University Press, Cambridge (1973)

Carter, B.: Black hole equilibrium states, In Black Holes—Les astres occlus. Gordon and Breach Science Publishers, (1973)

Hawking, S.W.: Gravitational radiation from colliding black holes. Phys. Rev. Lett. 26, 1344- 1346 (1971)

Hawking, S.: Black holes in general relativity. Commun. Math. Phys. 25, 152-166 (1972)

Bekenstein, J.: Extraction of energy and charge from a black hole. Phys. Rev. D7, 949-953 (1973)

Unruh, W., Wald, R.M.: Entropy bounds, acceleration radiation, and the generalized second law. Phys. Rev. D27, 2271-2276 (1983)

Image : MPI for gravitational physics / W.Benger-zib

Tomilin,K.A., (1999). "Natural Systems Of Units: To The Centenary Aniniversary Of The Planck Systems", 287-296

Sivaram, C. (2007). "What Is Special About the Planck Mass"? arXiv:0707.0058v1

Supernova Cosmology Project. Measurements of Ω and Λ from 42 high redshift supernovae. Astrophys. J. 517, 565 (1999).

Horndeski, G. W. Second-order scalar-tensor field equations in a four-dimensional space. Int. J. Theor. Phys. 10, 363 (1974).

Noller, J. & Nicola, A. Cosmological parameter constraints for Horndeski scalar-tensor gravity. Phys. Rev. D. 99, 103502 (2019).
Koyama, K. Cosmological tests ofmodified gravity. Rept. Prog. Phys. 79, 046902 (2016).

Song, Y.-S. & Percival, W. J. Reconstructing the history of structure formation using redshift distortions. JCAP 10, 004 (2009).

D. VALEV, Estimations of total mass and density of the observable universe by dimensional analysis, Aerospace Res. Bulgaria,24, pp. 67-76, 2012.

Determination of cosmological parameters, Astrophys. J. Suppl. Ser., 148, pp. 175-194, 2003.

Dirac's Large Number Hypothesis. Part 1, arXiv:gr-qc/0606057, 2006.

M. LACHIEZE-REY, E. GUNZIG, The cosmological background radiation, Cambridge Univ. Press, 1999.

E. GREGERSEN, The Universe: A historical survey of beliefs, theories and laws, Rosen Publ. Group, New York, 2010.

Glossary

Absolute space and time—the Newtonian concepts of space and time, in which space is independent of the material bodies within it, and time flows at the same rate throughout the universe without regard to the locations of different observers and their experience of "now."

Acceleration—the rate at which the speed or velocity of a body changes.

Accelerating universe—the discovery in 1998, through data from very distant supernovae, that the expansion of the universe in the wake of the big bang is not slowing down, but is actually speeding up at this point in its history; groups of astronomers in California and Australia independently discovered that the light from the supernovae appears dimmer than would be expected if the universe were slowing down.

Action—the mathematical expression used to describe a physical system by requiring only the knowledge of the initial and final states of the system; the values of the physical variables at all intermediate states are determined by minimizing the action.

Anthropic principle—the idea that our existence in the universe imposes constraints on its properties; an extreme version claims that we owe our existence to this principle.

Asymptotic freedom (or safety)—a property of quantum field theory in which the strength of the coupling between elementary particles vanishes with increasing energy and/or decreasing distance, such that the elementary particles approach free particles with no external forces acting on them; moreover for decreasing energy and/or increasing distance between the particles, the strength of the particle force increases indefinitely.

Baryon—a subatomic particle composed of three quarks, such as the proton and neutron.

Classical theory—a physical theory, such as Newton's gravity theory or Einstein's general relativity, that is concerned with the macroscopic universe, as opposed to theories concerning events at the submicroscopic level such as quantum mechanics and the standard model of particle physics.

Copernican revolution—the paradigm shift begun by Nicolaus Copernicus in the early sixteenth century, when he identified the sun, rather than the Earth, as the center of the known universe.

Cosmic microwave background (CMB)—the first significant evidence for the big bang theory; initially found in 1964 and studied further by NASA teams in 1989 and the early 2000s, the CMB is a smooth signature of microwaves everywhere in the sky, representing the "afterglow"of the big bang: Infrared light produced about 400,000 years after the big bang had redshifted through the stretching of spacetime during fourteen billion years of expansion to the microwave part of the electromagnetic spectrum, revealing a great deal of information about the early universe.

Cosmological constant—a mathematical term that Einstein inserted into his gravity field equations in 1917 to keep the universe static and eternal; although he later regretted this and called it his "biggest blunder," cosmologists today still use the cosmological constant, and some equate it with the mysterious dark energy.

Coupling constant—a term that indicates the strength of an interaction between particles or fields; electric charge and Newton's gravitational constant are coupling constants.

Curvature—the deviation from a Euclidean spacetime due to the warping of the geometry by massive bodies.

Dark energy—a mysterious form of energy that has been associated with negative pressure vacuum energy and Einstein's cosmological constant; it is hypothesized to explain the data on the accelerating expansion of the universe; according to the standard model, the dark energy, which is spread uniformlythroughout the universe, makes up about 70 percent of the total mass and energy content of the universe.

Dark matter—invisible, not-yet-detected, unknown particles of matter, representing about 30 percent of the total mass of matter according to the standard model; its presence is necessary if Newton's and Einstein's gravity theories are to fit data from galaxies, clusters of galaxies, and cosmology; together, darkmatter and dark energy mean that 96 percent of the matter and energy in the universe is invisible.

Dwarf galaxy—a small galaxy (containing several billion stars) orbiting a larger galaxy; the Milky Way has over a dozen dwarf galaxies as companions, including the Large Magellanic Cloud and Small Magellanic Cloud.

Dynamics—the physics of matter in motion.

Electromagnetism—the unified force of electricity and magnetism, discovered to be the same phenomenon by Michael Faraday and James Clerk Maxwell in the nineteenth century.

Electromagnetic radiation—a term for wave motion of electromagnetic fields which propagate with the speed of light—300,000 kilometers per second—and differ only in wavelength; this includes visible light, ultraviolet light, infrared radiation,

X-rays, gamma rays, and radio waves.

Eötvös experiments—torsion balance experiments performed by Hungarian Count Roland von Eötvös in the late nineteenth and early twentieth centuries that showed that inertial and gravitational mass were the same to one part in 1011; this was a more accurate determination of the equivalence principle than results achieved by Isaac Newton and, later, Friedrich Wilhelm Bessel.

Epicycle—in the Ptolemaic universe, a pattern of small circles traced out by a planet at the edge of its "deferent" as it orbited the Earth; this was how the Greeks accounted for the apparent retrograde motions of the planets.

Equivalence principle—the phenomenon first noted by Galileo that bodies falling in a gravitational field fall at the same rate, independent of their weight and composition; Einstein extended the principle to show that gravitation is identical (equivalent) to acceleration.

Escape velocity—the speed at which a body must travel in order to escape a strong gravitational field; rockets fired into orbits around the Earth have calculated escape velocities, as do galaxies at the periphery of galaxy clusters.

Ether (or aether)—a substance whose origins were in the Greek concept of "quintessence," the ether was the medium through which energy and matter moved, something more than a vacuum and less than air; in the late nineteenth century the Michelson-Morley experiment disproved the existence of the ether.

Euclidean geometry—plane geometry developed by the third-century bc Greek mathematician Euclid; in this geometry, parallel lines never meet.

Fermion—a particle with half-integer spin, like protons and electrons, that make up matter.

Field—a physical term describing the forces between massive bodies in gravity and electric charges in electromagnetism; Michael Faraday discovered the concept of field when studying magnetic conductors.

Field equations—differential equations describing the physical properties of interacting massive particles in gravity and electric charges in electromagnetism; Maxwell's equations for electromagnetism and Einstein's equations of gravity are prominent examples in physics.

Fifth force or **"skew" force**—a new force in MOG that has the effect of modifying gravity over limited length scales; it is carried by a particle with mass called the phion.

Frame of reference—the three spatial coordinates and one time coordinate that an observer uses to denote the position of a particle in space and time.

Galaxy—organized group of hundreds of billions of stars, such as our Milky Way.

Galaxy cluster—many galaxies held together by mutual gravity but not in as organized a fashion as stars within a single galaxy.

Galaxy rotation curve—a plot of the Doppler shift data recording the observed velocities of stars in galaxies; those stars at the periphery of giant spiral galaxies are observed to be going faster than they "should be" according to Newton's and Einstein's gravity theories.

General relativity—Einstein's revolutionary gravity theory, created in 1916 from a mathematical generalization of his theory of special relativity; it changed our concept of gravity from Newton's universal force to the warping of the geometry of spacetime in the presence of matter and energy.

Geodesic—the shortest path between two neighboring points, which is a straight line in Euclidian geometry, and a unique curved path in four-dimensional spacetime.

Globular cluster—a relatively small, dense system of up to millions of stars occurring commonly in galaxies.

Gravitational lensing—the bending of light by the curvature of spacetime; galaxies and clusters of galaxies act as lenses, distorting the images of distant bright galaxies or quasars as the light passes through or near them.

Gravitational mass—the active mass of a body that produces a gravitational force on other bodies.

Gravitational waves—ripples in the curvature of spacetime predicted by general relativity; although any accelerating body can produce gravitational radiation or waves, those that could be detected by experiments would be caused by cataclysmic cosmic events.

Graviton—the hypothetical smallest packet of gravitational energy, comparable to the photon for electromagnetic energy; the graviton has not yet been seen experimentally.

Inertia—the tendency of a body to remain in uniform motion once it is moving, and to stay at rest if it is at rest; Galileo discovered the law of inertia in the early seventeenth century.

Inertial mass—the mass of a body that resists an external force; since Newton, it has been known experimentally that inertial and gravitational mass are equal; Einstein used this equivalence of inertial and gravitational mass to postulate his equivalence principle, which was a cornerstone of his gravity theory.

Inflation theory—a theory proposed by Alan Guth and others to resolve the flatness, horizon, and homogeneity problems in the standard big bang model; the very early universe is pictured as expanding exponentially fast in a fraction of a second.

Interferometry—the use of two or more telescopes, which in combination create a receiver in effect as large as the distance between them; radio astronomy in particular makes use of interferometry.

Inverse square law—discovered by Newton, based on earlier work by Kepler, this law states that the force of gravity between two massive bodies or point particles decreases as the inverse square of the distance between them.

Metric tensor—mathematical symmetric tensor coefficients that determine the infinitesimal distance between two points in spacetime; in effect the metric tensor distinguishes between Euclidean and non-Euclidean geometry.

Michelson-Morley experiment—1887 experiment by Albert Michelson and Edward Morley that proved that the ether did not exist; beams of light traveling in the same direction, and in the perpendicular direction, as the supposed ether showed no difference in speed or arrival time at their destination.

Milky Way—the spiral galaxy that contains our solar system.

Minkowski spacetime—the geometrically flat spacetime, with no gravitational effects, first described by the Swiss mathematician Hermann Minkowski; it became the setting of Einstein's theory of gravity.

MOG—my relativistic modified theory of gravitation, which generalizes Einstein's general relativity; MOG stands for "Modified Gravity."

MOND—a modification of Newtonian gravity published by Mordehai Milgrom in 1983; this is a nonrelativistic phenomenological model used to describe rotational velocity curves of galaxies; MOND stands for "Modified Newtonian Dynamics."

Newton's gravitational constant—the constant of proportionality, G, which occurs in the Newtonian law of gravitation, and says that the attractive force between two bodies is proportional to the product of their masses and inversely proportional to the square of the distance between them; its numerical value is: G = 6.67428 ± 0.00067 x 10-11 m3 kg-1 s-2.

Nonsymmetric field theory (Einstein)—a mathematical description of the geometry of spacetime based on a metric tensor that has both a symmetric part and an antisymmetric part; Einstein used this geometry to formulate a unified field

theory of gravitation and electromagnetism.

Nonsymmetric Gravitation Theory (NGT)—my generalization of Einstein's purely gravitation theory (general relativity) that introduces the antisymmetric field as an extra component of the gravitational field; mathematically speaking, the nonsymmetric field structure is described by a non-Riemannian geometry.

Parallax—the apparent movement of a nearer object relative to a distant background when one views the object from two different positions; used with triangulation for measuring distances in astronomy.

Paradigm shift—a revolutionary change in belief, popularized by the philosopher Thomas Kuhn, in which the majority of scientists in a given field discard a traditional theory of nature in favor of a new one that passes the tests of experiment and observation; Darwin's theory of natural selection, Newton's gravity theory, and Einstein's general relativity all represented paradigm shifts.

Parsec—a unit of astronomical length equal to 3.262 light years.

Perihelion—the position in a planet's elliptical orbit when it is closest to the sun.

Perihelion advance—the movement, or changes, in the position of a planet's perihelion in successive revolutions of its orbit over time; the most dramatic perihelion advance is Mercury's, whose orbit traces a rosette pattern.

Phion—name given to the massive vector field in MOG; it is represented both by a boson particle, which carries the fifth force, and a field.

Planck's constant (h)—a fundamental constant that plays a crucial role in quantum mechanics, determining the size of quantum packages of energy such as the photon; it is named after Max Planck, a founder of quantum mechanics

Principle of general covariance—Einstein's principle that the laws of physics remain the same whatever the frame of reference an observer uses to measure physical quantities.

Quantize—to apply the principles of quantum mechanics to the behavior of matter and energy (such as the electromagnetic or gravitational field energy); breaking down a field into its smallest units or packets of energy.

Quantum field theory—the modern relativistic version of quantum mechanics used to describe the physics of elementary particles; it can also be used in nonrelativistic fieldlike systems in condensed matter physics.

Quantum gravity—attempts to unify gravity with quantum mechanics.

Quantum mechanics—the theory of the interaction between quanta (radiation) and matter; the effects of quantum mechanics become observable at the submicroscopic distance scales of atomic and particle physics, but macroscopic quantum effects can be seen in the phenomenon of quantum entanglement.

Quantum spin—the intrinsic quantum angular momentum of an elementary particle; this is in contrast to the classical orbital angular momentum of a body rotating about a point in space.

Quintessence—a fifth element in the ancient Greek worldview, along with earth, water, fire and air, whose purpose was to move the crystalline spheres that supported the heavenly bodies orbiting the Earth; eventually this concept became known as the "ether," which provided the *something* that bodies needed to be in contact with in order to move; although Einstein's special theory of relativity dispensed with the ether, recent explanations of the acceleration of the universe call the varying negative pressure vacuum energy "quintessence."

Riemann curvature tensor—a mathematical term that specifies the curvature of four-dimensional spacetime.

Riemannian geometry—a non-Euclidean geometry developed in the mid-nineteenth century by the German mathematician George Bernhard Riemann that describes curved surfaces on which parallel lines *can* converge, diverge, and even intersect, unlike Euclidean geometry; Einstein made Riemannian geometry the mathematical formalism of general relativity.

Scalar field—a physical term that associates a value without direction to every point in space, such as temperature, density, and pressure; this is in contrast to a vector field, which has a direction in space; in Newtonian physics or in electrostatics, the potential energy is a scalar field and its gradient is the vector force field; in quantum field theory, a scalar field describes a boson particle with spin zero.

Scale invariance—distribution of objects or patterns such that the same shapes and distributions remain if one increases or decreases the size of the length scales or space in which the objects are observed; a common example of scale invariance

is fractal patterns.

Schwarzschild solution—an exact spherically symmetric static solution of Einstein's field equations in general relativity, worked out by the astronomer Karl Schwarzschild in 1916, which predicted the existence of black holes.

Self-gravitating system—a group of objects or astrophysical bodies held together by mutual gravitation, such as a cluster of galaxies; this is in contrast to a "bound system" like our solar system, in which bodies are mainly attracted to and revolve around a central mass.

Singularity—a place where the solutions of differential equations break down; a spacetime singularity is a position in space where quantities used to determine the gravitational field become infinite; such quantities include the curvature of spacetime and the density of matter.

Spacetime—in relativity theory, a combination of the three dimensions of space with time into a four-dimensional geometry; first introduced into relativity by Hermann Minkowski in 1908.

Special theory of relativity—Einstein's initial theory of relativity, published in 1905, in which he explored the "special" case of transforming the laws of physics from one uniformly moving frame of reference to another; the equations

of special relativity revealed that the speed of light is a constant, that objects appear contracted in the direction of motion when moving at close to the speed of light, and that E = mc2, or energy is equal to mass times the speed of light squared.

Tully-Fisher law—a relation stating that the asymptotically flat rotational velocity of a star in a galaxy, raised to the fourth power, is proportional to the mass or luminosity of the galaxy.

Unified theory (or unified field theory)—a theory that unites the forces of nature; in Einstein's day those forces consisted of electromagnetism and gravity; today the weak and strong nuclear forces must also be taken into account, and perhaps someday MOG's fifth force or skew force will be included; no one has yet discovered a successful unified theory.

Vacuum—in quantum mechanics, the lowest energy state, which corresponds to the vacuum state of particle physics; the vacuum in modern quantum field theory is the state of perfect balance of creation and annihilation of particles and antiparticles.

Variable Speed of Light (VSL) cosmology—an alternative to inflation theory, in which the speed of light was much faster at the beginning of the universe than it is today; like inflation, this theory solves the horizon and flatness problems in the very early universe in the standard big bang model.

Vector field—a physical value that assigns a field with the position and direction of a vector in space; it describes the force field of gravity or the electric and magnetic force fields in James Clerk Maxwell's field equations.

Virial theorem—a means of estimating the average speed of galaxies within galaxy clusters from their estimated average kinetic and potential energies.

www.ingramcontent.com/pod-product-compliance
Lightning Source LLC
LaVergne TN
LVHW022320200926
848481LV00004B/1381